Time Travel

Fun Facts & Theories on How to Travel Through Time & Space

KIV Books

Copyright © 2017

Copyright © 2017 KIV Books

Disclaimer

This book is designed to provide condensed information. It is not intended to reprint all the information that is otherwise available, but instead to complement, amplify and supplement other texts. You are urged to read all the available material, learn as much as possible and tailor the information to your individual needs.

Every effort has been made to make this book as complete and as accurate as possible. However, there may be mistakes, both typographical and in content. Therefore, this text should be used only as a general guide and not as the ultimate source of information. The purpose of this book is to educate.

The author or the publisher shall have neither liability nor responsibility to any person or entity regarding any loss or damage caused, or alleged to have been caused, directly or indirectly, by the information contained in this book.

Contents

Introduction

Before you start this book ask yourself these questions:
What do you know about Traveling Through Time & Space?
Are you keen to learn about it?

This book will tell you about Space and Time and it will explain vital terms in a simple yet factual way that does not blind the reader with science.

It will tell you about the vitally important topics of the Solar System, Flight, and Light.

It will introduce you to the fascinating topics of Black Holes, Dark Matter, and Wormholes, which are so important to anyone with an interest in Space & Time.

The book finishes with Chapters about UFOs, both Modern and Ancient, who we believe carry those who have already mastered Space & Time, and a light-hearted look at those who see them as a Threat.

Onto Chapter 1.

Space, Stars, Planets, Time

"Astronomy compels the soul to look upward and leads us from this world to another."
— Plato, The Republic, 342 BCE.

Space: When space is mentioned many people think of stars and planets. These heavenly bodies fascinated ancient man. Their movement across the void was useful but mysterious. Calendars were based on observations of the Sun. It was thought they must be controlled by something supernatural. Many believed in astrology, which is the belief that the positions of those heavenly bodies affect our lives. Indeed many still do.

Centuries passed with no distinction made between astrology and astronomy, which is the scientific study of the heavens. Some beliefs of the early astronomers were just as misleading as the prognostications of the astrologers. As an example, the Earth was believed to be at the center of the Universe. It is only in the last few hundred years that we have come to understand that Earth is a tiny dot in space. This small planet goes around the Sun which is just one of about hundred billion stars in our galaxy and there are at least a billion other galaxies in the universe.

Stars: We have all seen stars. They are visible on clear nights as very small points of light in the sky that sparkle. Stars feature in a vast body of literature, poem, and melody.

They are incandescent balls of gas, mainly the gasses helium and hydrogen. Their enormous mass produces the gravity, which holds them together. There are nuclear fusion reactions in their core. They produce heat and photons, as well as small amounts of heavier elements.

The closest star to Earth is, of course, the Sun. The distance of this familiar star from Earth is about 90 Million miles or 150 million kilometers. The star that is closest next is Proxima Centauri. It is a vast 5.878 trillion miles or 9.461 trillion kilometers away!

There are about 6000 stars in the night sky that can be seen with the eye without the aid of instruments like telescopes. Not all these stars can be seen at once, as only a small part of the heavens is visible from any single spot on Earth.

The brightest stars are visible by themselves, but others are more easily seen as part of constellations. Constellations are groups of stars that form patterns. They were well known to ancient people who thought that the stars were patterned in the forms of animals, mythical creatures, and human beings. There are 88 constellations.

Planets: It has been agreed by scientists that an object in space is a planet if it fulfills the following:

1. It moves in orbit around a star

2. It is sufficiently large that it assumes a spherical shape as result of its own gravitational forces

3. It does not interfere with anything else in the neighborhood of its orbit

The Greek word for *wanderer* is close to *our word for a planet*. The reason for this is that the planets seem to wander among the other stars, which seem to be fixed in the heavens. By the middle of 2013, about 900 planets had been discovered. More have been found since then. Scientists believe that it is very likely that there are trillions and many of these trillions of planets will be able to have life, as we know it.

Time: We all think we know what time is. If someone says, "What is the time?" We look at our watch but it gets more complicated. The best scientists, astronomers, and mathematicians don't know what time is. Einstein in his theory of relativity treated time as the fourth dimension yet we know that time is different to the other side dimensions. Time cannot be measured with rulers or any other similar yardstick, and furthermore, although we can be at one unchanging point in space using the other three dimensions time is always changing!

Scientists and Astronomers

"Truth searches for no one. It waits to be found."
— Suzy Kassem, Rise Up and Salute the Sun: The Writings of Suzy
Kassem

What is a scientist? A scientist is a person who engages in an organized activity in order to obtain knowledge that can describe and make predictions about the natural world. Scientists are different to engineers. Engineers design, build and maintain things for human beings however most engineers need to study science before they can do their jobs.

What is an astronomer? Astronomers study space and heavenly bodies. It is the oldest science of all as this study was vital to early man for the establishment of calendars, allowing a sensible scheduling of activities. The explanations provided for normal activity such as the rising and setting of the Sun, the waxing, and waning of the Moon etc. were not, by our standards, very scientific, though, often being seen as under the control of the gods. In earlier days most astronomers believed that the world was at the center of the universe.

This continued until the 16th century when both calculation and observation, with the newly invented telescope, revealed that the Earth rotated around the Sun rather than the Sun around the Earth. A century later the great English mathematician, Isaac Newton, discovered gravity and towards the end of the 18th century another British astronomer called William Herschel worked out the shape of the Milky Way, which is our galaxy. He also proposed that there were other galaxies at great distances from us, other than our own.

Early in the 20th century Einstein, with his theory of general relativity, gave astronomy a huge theoretical boost. Observations by the American astronomer Edwin Hubble showed that the universe is expanding. Hubble also confirmed the proposition of Herschel that there were other galaxies. In fact, there are billions of them.

Our Solar System and Galaxy

"In more than one respect, the exploring of the Solar System and homesteading other worlds constitutes the beginning, much more than the end, of history."
— Carl Sagan

Our solar system consists of the Sun, nine planets, their moons and a few other things, which will be mentioned towards the end of this chapter.

The Sun: The Sun by itself is greater than 750 times as much as the rest of the Solar System combined. Its huge gravity locks everything within a range of over 6 billion kilometers into an orbit around it. After the Sun, we have the nine planets, but let us first consider the Sun. Our sun is a yellow dwarf, something we will talk much more of in a later chapter. This means it is a small, dim star. The reason it seems so large and bright is that it is much closer than any other star, only 150 million kilometers (93 million miles) away from Earth.

As with all stars, the sun emits energy produced by nuclear re-actions in its core. There, the temperature is around 15 million degrees Celsius compared with about 6,000 degrees on its surface. In these reactions, around 600 million tons of hydrogen are converted to helium each second. This has been going on for around 5 billion years, there is enough hydrogen left for the sun to continue like this for another 5 billion years,

The sun radiates this energy as *electromagnetic Waves*. Among these are heat, light, radio waves, X-rays, ultraviolet rays, and gamma rays. Only a fraction of this energy reaches the earth. As a result, we get the light, heat and, indirectly, almost all the energy we use.

Planets closest to the Sun: These planets are Mercury, Venus, Earth and Mars. Apart from Earth, they have barren and rocky terrain. Earth is different because of the seas and oceans that cover most of it. Mercury and Mars have very thin atmospheres causing a great temperature difference between day and night. Earth and Venus have shielding atmospheres. Their temperatures are fairly constant. The temperature on Earth is ideal for life but on Venus, the temperature is far too hot for life, as we know it.

Because Mercury and Venus are so close to the sun, both planets have scorching daytime temperatures. However, they have little else in common. The diameter of Venus is about two and a half times that of Mercury. Hence its volume is around 16 times as great.

Mercury has almost no atmosphere and looks like the Moon. A vast, dense atmosphere that shields our view of its surface covers Venus. The two planets are different in the direction of rotation. Venus turns in an opposite direction to most planets. Such a spin is called *retrograde.*

Mercury: Mercury is the smallest of the main planets and is not much larger than our moon. Like the Moon, it shows phases. Mercury resembles the Moon in other ways too. Its surface is pitted with craters. Its temperature changes greatly from day to night. During the day it reaches a maximum of about 400°C. At night, the temperature drops to around -200'C that is cold enough to freeze oxygen gas into a liquid. The nights are extremely cold because there is no atmospheric blanket to retain the heat.

Venus: Venus is easily seen from the Earth. Sometimes it is called the Morning Star or the Evening Star. It looks like a bright star and shows in the East at sunrise, or in the West at sunset. Venus looks so bright because it has a dense cover of white clouds. These reflect sunlight well and make Venus brighter than any star in the sky except of course the Sun.

The atmosphere around Venus consists mainly of carbon dioxide gas. The dense clouds are probably made up of tiny droplets of sulfuric acid. The temperatures on Venus during the day reach about 500°C. This heat is kept in by the atmosphere and keeps the planet relatively warm at night.

Little is known about the surface of Venus because it is covered by cloud. However, some craters have been detected by radar. The Soviet Union (Russia) sent some probes there decades ago but not much was discovered as a result of the harsh conditions.

Earth: No one knows for certain how the Earth was formed. Scientists have discovered a lot about how it developed from a ball of molten rock to its current state. From space, the Earth is a marvelous, bright globe.

About three-quarters of its surface are covered with water, which reflects sunlight well. Between the brilliant white polar ice caps are blue oceans and brownish continents, sometimes covered with patches of white clouds

About 4.6 billion years ago Earth was formed. At that time it was a molten mass of very hot rock. Over many millions of years, the rock gradually cooled, and a solid crust formed. The crust often cracked while it cooled down. Molten rock from the interior then oozed through the cracks and solidified at the surface.

As cooling occurred steam and other gasses escaped from the inside. Most of the steam then condensed to form the seas, oceans, lakes etc. The other gasses stayed above the surface as the atmosphere, mainly of nitrogen and oxygen. The gravity of the Earth stopped the atmosphere from escaping into space.

About a billion years ago the surface of Earth was one huge island called Gondwanaland surrounded by a vast ocean. Stresses in the crust of the Earth slowly pulled the island asunder to form separate continents. The East coast of South America and the West coast of Africa are very similar in shape as they were once joined. The shaping of the surface took place quite late in the history of the Earth.

About 400 million years ago, movements in the crust caused parts of it to fold and form the mountain chains. Some mountains were formed more recently. The Himalayas, the Alps, and the Rocky Mountains have formed only 25 million years ago.

The Earth seems relatively stable, but it is still slowly changing and the gradual separation of enormous parts of the crust continues. Sudden slipping of one plate (section) against another causes earthquakes. In some places where the crust is weak or thin molten rock from below bursts through the surface causing volcanoes. The stream of red-hot rock that pours out of the volcano is called *lava*.

No other planets in the Solar System can support the various complex forms of life that exist on Earth. Only Earth has the conditions necessary for living organisms to develop to such an extent. We do not know how life began, however; it is very unlikely that there could have been life without water.

The first, tiny, primitive organisms may have developed over 3 billion years ago. Over millions of years, numerous much more complex creatures developed. Scientists believe that this slow process, called *evolution*, led to the great variety of plants and animals found on the Earth today.

Mankind appeared on the scene quite recently on the Earth. He appeared less than three million years ago. Modern man only appeared 200 000 years ago.

Mars: Mars has always had a special attraction for astronomers. It is the planet most like Earth. It is the only one whose surface markings can be seen from the earth. The other planets either have a dense cloud cover or are too distant to see clearly. When conditions are good a small telescope will reveal the red Martian surface with dark markings and white polar caps.

 Early observers found many comparisons between features on Mars and on Earth. They thought that the Martian polar caps consisted of frozen water and that its yellow regions were great sand-covered deserts. They thought the dark surface markings were seas and oceans.

With so many similarities between Mars and Earth, the existence of life on Mars seemed quite possible. This idea received a boost in 1877 when 'canals' were discovered on Mars. A lot of astronomers believed these straight markings were irrigation systems built by Martians to irrigate the desert regions. Later and better observations showed that this and many other ideas about Mars were quite wrong.

There have been a number of unmanned missions to Mars but we may have to wait until man manages to get there and bring samples of Martian soil back before we know, for certain, whether or not life exists on Mars.

The Outer Planets: Beyond Mars lies the Asteroid Belt. We will talk more about asteroids later in this chapter. Beyond the Asteroid Belt are the outer planets. These planets are Jupiter, Saturn, Uranus, and Neptune. They are vast balls of gas orbiting in the Solar Systems. Beyond these big planets lies tiny frozen Pluto.

Jupiter: Jupiter is the Goliath of the Solar System. It has a volume more than 1000 times that of Earth. It does not have a solid surface. The upper layers of Jupiter are a mix of gasses that become liquid and finally solid near the center. In the cloud layer the temperatures average -140°C.

On the surface of Jupiter is something called the Great Red Spot. It was first observed, with a small telescope, in 1666. It is thought by scientists that it may be a long-lasting storm raging in the atmosphere of the planet. There have been a number of probes around Jupiter. They have discovered that there is a faint ring around the planet, there are two more moons than before and also on its surface, there are some huge volcanoes.

In 1992 the comet Levy-2 collided with Jupiter creating some spectacular sights. In March 2016, Jupiter was hit by what was probably an asteroid

Saturn: Saturn orbits at a distance of about 1.4 billion km from the sun. Like Jupiter, it is a gas giant. What makes it truly amazing and interesting is that it has rings which surround it. Seen from Earth the rings appear different from year-to-year as the planet moves in its orbit. The rings are about 1 km (half a mile) thick, and not solid.

The rings are composed of many small pieces of rock and ice. In 1980, the probe Voyager 1 discovered that the rings are divided into hundreds of distinct 'ringlets'. The two outermost ringlets are twisted around each other like the strands of a rope.

Saturn has at least 15 moons, most of them made of ice and each of them cratered by many comet impacts. One has a crater, which is a quarter of its own size, and another has been smashed into two smaller moons, which now share the same orbit.

The largest moon, Titan, has a thick atmosphere of nitrogen, and orange clouds, which completely hide its service. The temperature of Titan is about -180°C. Some astronomers think oceans of liquid methane cover it. On Earth, methane is a fuel.

Uranus: Uranus is a very cold planet, about 2.9 billion km from the Sun. It has five moons and it has also got nine narrow rings, much darker than those of Saturn. A probe has not visited it since Voyager 2 in 1986.

Neptune: Neptune is the last of the giant planets. It is similar to Uranus but is slightly smaller in size. It has two moons, which are named Triton and Nereid. It has cloud layers, which have been calculated to be at a temperature of -220°C. Voyager 2 visited it in 1989. Voyager 2 passed within 5000 km or 3000 miles of Neptune.

Pluto: Pluto is the second smallest of the known planets and has a diameter of only 6400 km (3840 miles). The distance of Pluto from the sun is greater than that of any other planets. The orbit of Pluto is off-center and crosses the orbits of Neptune. Little is known about Pluto but it is probably composed of extremely dense material.

Some astronomers believe that Pluto was once a moon of Neptune. In 1972 some American astronomers predicted that another planet would be found beyond Pluto at a distance of about 9.6 trillion km from the sun.

The gravitational attraction of such a planet would account for slight variations in the orbits of some comets. The existence of

Pluto was predicted 15 years before this discovery in a similar way.

Pluto has one moon called Charon. Charon is about half the size of Pluto. Pluto was visited by the New Horizons probe in 2015. New Horizons passed within 12500 km or 7800 miles of Pluto. Pluto had its status lowered from planet to dwarf planet in 2006.

This brief summary covers the planets but there are other celestial objects in the Solar System.

Asteroids: Asteroids are rocks in space. They were probably formed at the same time as the planets. Most asteroids orbit the Sun in a nearly circular path between Mars and Jupiter called the Asteroid Belt. This was mentioned before. Some asteroids have eccentric orbits.

Few asteroids are greater than 160 km or 100 miles in diameter. If one of the larger asteroids were to collide with the Earth it would be absolutely disastrous and probably wipe out most life on the Earth. Such a collision in Mexico is thought to have exterminated the dinosaurs 66 million years ago.

Comets: Comets are regular visitors to our part of the solar system. They are composed of dust, ice, and frozen gasses. Comets glow and develop a tail of glowing gasses when they pass near the sun. Like planets, comets orbit the Sun. The orbits of most planets are almost circular; the orbits of comets are very long and extremely eccentric.

Although comets regularly pass near the sun, they spend most time among, or beyond, the orbits of the planets. Some comets have orbits, which are completely within our planetary system. They are called *short-term* comets. The most famous of these is

Halley's Comet, which never quite reaches the orbit of Pluto. Halley's Comet passes close to the sun every 76 years and can be seen with the eye at this time. The last time this occurred was in 1986. *Long-term comets* travel beyond Pluto and are rarely seen.

There are estimated to be about 100 billion comets altogether but many of them pass close to the sun only once in several million years. In most years, about five comets are visible from the Earth. Normally, they are too dim to be seen with the naked eye. Occasionally though they are quite spectacular. The brightest of them are visible in broad daylight, with tails going halfway across the sky. In past ages, they were thought of as portents of doom.

Meteoroids and meteors: Meteoroids are solid bodies that have broken away from comets or asteroids. The smallest meteoroids are no bigger than grains of sand, while the largest meteoroids may weigh several tons.

Meteoroids are always entering the atmosphere of the Earth. The friction with the atmosphere normally heats them up to such a degree that they burn up before reaching the ground. This gives rise to sudden bright streaks called meteors, shooting stars, or falling stars, in the night sky. If you are outside on a clear night you will almost certainly see some, as well as satellites.

A meteor shower is a spectacular meteor display that happens when the Earth passes through a cloud of meteoroids, resulting in many, sometimes thousands of, meteors within a short time.

Large meteoroids occasionally survive the journey through the atmosphere and smash into the ground. We call these rocks from outer space meteorites. Most meteorites are small rocks but the largest may be huge and blast out huge craters when they collide with the ground. This, fortunately, is very rare.

There was a recent case, where this actually did occur in Russia in 2013. Google, ' Meteorite hitting Earth in Russia YouTube.'

Galaxy: Galaxies are immense star systems. They are the building blocks of the Universe. Billions of galaxies exist.

The Sun is just one of about 100 billion stars in a vast, rotating spiral arrangement that is referred to as our Galaxy and given the name Milky Way. On a clear night, it is possible to see thousands of these stars with the naked eye. In some parts of the sky, there are so many stars that they appear to form white bands.

The distance across our Galaxy is about 1 million trillion km. Astronomers usually express large distances like this in terms of *light years* which are the distance that light travels through space in one year.

Besides stars, the arms of the spirals of the Galaxy contain clouds of hydrogen gas and also *nebulae*, vast clouds of dust and gasses. It is thought that new stars are created from the nebulae.

Flight

"Once you have tasted flight, you will forever walk the earth with your eyes turned skyward, for there you have been, and there you will always long to return."
— Leonardo da Vinci

The gods of the ancients lived in the heavens. Many of them flew. In Egypt Isis was a winged goddess who flew in the heavens. In Sumeria, the god of the sun was Samas who flew daily across the skies. In Mexico, Quetzalcoatl was a mythological winged serpent that created mankind. It has always been the wish of mankind to fly.

What follows is a brief history of mankind's flying.

The history of aviation goes back for more than 2000 years. Kite flying started in China at some date hundred of years BC and gradually spread to the rest of the world. The flying of kites is probably the earliest man-made flight. Leonardo da Vinci made drawings of both planes and helicopters. It is doubtful if he ever made any, though.

The next major steps in the development of flight were balloons in the 18th century and gliders in the 19th century. Finally, at the beginning of the 20th century, there was manned flight.

Probably the first person to do this was Richard Pearse, a New Zealander, in 1903. His flight was about nine months before that of the Wright brothers who are officially credited with the first manned flight.

There were other people in a number of countries who succeeded in flying to a greater or lesser degree at this time. This is a good example of something that was bound to happen, as the technology was ready. It was a question of when and not if.

There are other ways of flying such as blimps, helicopters, and hovercraft but we won't consider them.

To go into space you need something else: balloons, winged aircraft, blimps etc. require air in order to fly. As there is no air in the space another method has to be used in order to go there.

Rockets: The rockets of today are one of the most extraordinary achievements of humanity. They are the result of thousands of years of work and development. They are the only way we currently have to go into space.

When did rockets appear? By 100 A.D., there was in China a basic form of gunpowder. They used the explosions from this powder for ceremonies and from this simple start the rocket was born.

The first military use saw the Chinese attach these rockets to arrows fired by archers. It was found arrows were unnecessary. Rockets were on their way.

In the early part of the thirteenth century, there was a war between the Chinese and the Mongols. During one battle, the Chinese defeated the Mongols with a lot of these simple rockets.

As is usual with weapon systems, after this battle the Mongols made rockets of their own and probably Europeans first learned about rockets from them. Throughout the next two centuries, there were developments in rocketry in a number of European countries, including France, Italy, and England.

During the final part of the seventeenth century, the scientific foundations for modern rocketry were established with Newton's Three Laws of Motion. These laws allowed rockets to be developed systematically.

They were developed throughout Europe and were used in war. The British used them in the 1812 war against the United States. The rockets were very successful in that war.

After that though and up till the end of World War 1 rockets did not play a major role in the war as high-power artillery became available. This was far more devastating at that time than rockets.

In 1898 in Russia, a prescient visionary called Tsiolkovsky suggested exploring space with rockets. A few years later he came up with the idea of using liquid fuel in rockets so that the distance traveled would be increased.

In the early years of the twentieth century, rockets fascinated a brilliant American called Robert Goddard. He did a huge amount of work on them and enormously improved the rockets, which were in existence then.

During the Second World War, rockets came into their own. The Germans were shocked in 1941 when the Russians brought in the Katyusha rocket system called *Stalin organs*. These were devastating; later on, the Germans used such weapons themselves. The Americans used similar rocket systems in their island-hopping campaigns in the Pacific.

The Germans developed a very good rocket called the V2 near the end of World War Two. Unfortunately for them, it came into the war too late to affect the course of the Second World War.

The leader of the program, which built them, Werner von Braun, led the space program of the Americans during the 1950s and 1960s. The Russians had also spent a lot of time and money in developing a space program and on October 4, 1957, they launched the first artificial satellite. It was called Sputnik 1.

This event shocked the Americans who made massive changes to their educational and other systems as a consequence. It also started the space race, which for a while the Russians appeared to be winning. Following their success with the Sputnik, they were the first country to have a spaceman orbit the Earth. Yuri Gagarin did this in 1961. It was not until 1962 that John Glenn of the USA also did this.

The next step would be landing on the Moon; Russian and American scientists sent many probes to the Moon in the early 1960s. By 1969, the Americans were ready to land on the Moon. On the morning of July 21, 1969, astronauts Armstrong and Aldrin stepped onto the lunar surface.

The Americans followed this initial voyage to the Moon with five more visits, the last in 1972. The Russians never landed a person on the Moon. Once the Americans had done this the emphasis of both countries turned to other things like building space stations and space shuttles.

There is one other feature of the use of rockets that should be mentioned and that is their use in the arms race between the USA and the USSR. Both countries invested massively in ever more powerful weapon systems featuring rockets. Such rockets were called ICBMs (intercontinental ballistic missiles).

In the early 1980s, the Russians began to deploy SS-20 tactical missiles in Eastern Europe. The response of the US was to deploy Pershing 2 missiles in Europe and cruise missiles in Britain. Cruise missiles are modeled on the V1 rockets that the Nazis bombed Britain in 1944, killing thousands.

Shortly afterward Russia (USSR) experienced severe economic problems; Mikhail Gorbachev came to power there and together with Ronald Reagan, who was President of the USA, began the process of scaling this arsenal back. In the last couple of years, tensions between these two great countries have again risen.

Satellites: Sputnik 1, the first man-made satellite, was about the size of a large ball, 58 cm or 23 inches in diameter. It weighed only 83.6 kg or 183.9 pounds and took 98 minutes to orbit the Earth in an elliptical path. All it did was emit beeps.

Since then 8000 satellites have been put into orbit. At the moment 3000 are still in an orbit. Of this 3000 there are 943 operational satellites and of these 438 are American.

The largest satellite ever launched is an American spy satellite, which was launched in 2010. The largest commercial satellite launched was the Terrestar-1 launched in 2009. It has a wingspan of 32.4 m or 106 feet. Like most satellites launched for commercial reasons, it is a communications satellite.

58 countries have satellites but only 10 have launched their own. The European Union also has satellites. These are launched with Ariane rockets.

Will rockets take us into deep space? The reader will see that although the rocket has been a brilliant tool for what we have done something else will be needed in order for people to go much further than the moon, and possibly the planets.

How fast do space rockets go? At the beginning of the space age, scientists in Russia used the term *cosmic velocities* for some velocities that are critical in space exploration.

The first cosmic velocity also called the orbital velocity takes a rocket or other vehicle into orbit about the Earth. If the projectile is slower then it falls backs to Earth like an expensive firecracker. The second cosmic velocity must be reached if the rocket is to escape from Earth. This is 11.2 kilometers per second or 6.72 miles per second.

If a rocket goes fast enough it is possible to leave the Solar System and in order to do so, you need the third cosmic velocity. This is about 42 kilometers per second or 25 miles per second.

If you had a spacecraft capable of intergalactic travel then there is a fourth cosmic velocity, which is the velocity, required to be reached before you can escape our Galaxy. It is 320 kilometers a second or 192 miles a second.

Are any trips to planets planned? President Obama said this would be done during the 2030s with much work being done now. He hoped for a collaborative effort with other countries, particularly Russia. In a time of growing tension between the USA and Russia, this may not happen.

There are private organizations, which are planning one-way trips to Mars in the 2020s. The idea would be to establish colonies there. It remains to be seen whether this occurs.

How Long Would It Take to Go to Mars? The total traveling time from Earth to Mars takes 150-300 days depending on the launch speed, how Earth and Mars are aligned, and the length of the journey the spacecraft must undertake.

Journey time is very dependent on fuel. The more fuel you have, the faster you can go, and the shorter the travel time.

Light and Its Speed

"There are darknesses in life and there are lights, and you are one of the lights, the light of all lights."
— Bram Stoker

What is light? If you asked a reasonably intelligent person this they would say that light is the thing that enables us to see in the dark. They are of course correct but that answer is not going to be very helpful to us in our investigations of space and time.

A more useful definition is that light is an electromagnetic vibration and is very similar to heat, radio, infrared and other forms of radiation. As a form of radiation, light is comprised of *photons.*

A photon is a particle, which represents a very small quantity (quantum) of light or other electromagnetic radiation. A photon has zero rest mass but it has energy proportional to the radiation frequency of the particular electromagnetic radiation.

What is the speed of light? The speed of light is 300,000 kilometers per second or 186,000 miles per second. One hundred thousand light years would be needed in order to cross our galaxy at this speed. The distance across the Galaxy is 100,000 light years. The sun is a tiny dot 30,000 light years from the center of the vast disc of stars, which is the Milk Way.

The two sections above provide some almost unbelievable consequences. One of these is that there will be far more to space travel than we might think. Another is that light is closely associated with time travel!

Einstein called this speed the speed limit of the Universe. He said that a speed that was faster than that of light would violate the principle of cause and effect. An example of this would be a bullet, which hits a target before the trigger was pulled.

Going beyond it violates energy conditions. It even allows for time travel. Photons (light 'particles') travel at that speed because they have no mass. Things that have mass must go slower.

Before we go on it is necessary to introduce something called the *Higgs Field*. This is in space; indeed it is everywhere. Every particle in our universe, including photons, moves through it.

Particles gain their mass as they move through this field. Some particles have greater mass than others because of the way they interact with it. Photons move through the field but do not interact with it at all.

Particles get heavier as they go faster: Particles that have mass require energy to accelerate them, which means to make them go faster.

An electron has mass, although not much, certainly a lot less than you. Suppose you wanted to accelerate one of these tiny particles to the speed of light then you would need an infinite amount of energy. It will become infinitely heavy; there is not enough energy in the entire universe to propel the particle to this speed limit.

It is an extraordinary fact that as you drive along a road in a car and accelerate the car then you become minutely heavier. At the normal speeds at which we drive the increase would be so small, even if you were going as fast as a jet fighter, as to be negligible but would rapidly be noticeable as you got near the speed of light.

Relative speed is different near the speed of light:

If you are traveling along at 20 miles an hour and someone else passes you, going in the same direction, at 60 miles per hour then the speed of the passer relative to you is 60 - 20 = 40 miles an hour.

However, if that person passed you at the speed of light then it is a remarkable fact that you would measure his/her speed relative to you to be the speed of light. Also, you could never catch up as in order to do so you would have to go faster than light, which is impossible.

What we reveal next is in some ways frightening.

The effects of speed on time nearing the speed of light: Light years are a measurement of the distance light travels in one year.

Things are different if you're traveling near the speed of light than at the speeds at which we move on Earth. If you were moving at 10% the speed at which light travels, and you left Earth for some point one light year away an observer staying on Earth and watching you travel will measure ten years for you to get to the point.

However, as you're moving very fast, it turns out that the observer sees time passing more slowly than you do. From your point of view (onboard the spacecraft), it requires less than ten years to get to that point! In this situation, there is only a small difference: with the correct calculations, it will be found that it takes you about 0.994 of ten years. It seems to you that your trip as being about twenty days shorter than for the observer.

This effect becomes greater and greater as you speed up. If you move at 0.99999999 of the speed that light moves then from the observer's view, you're more or less moving at this speed and it will take you one year to get to the point.

However, for you, the time which passes is a little bit more than an hour. This is caused by *relativistic time dilation*. This extraordinary phenomenon is illustrated more dramatically by the following example.

Suppose you, age 32, are married to a lovely wife, age 32, and have three gorgeous children aged 8, 6 and 4. You are assigned the task of going to a star, which is 20 light years away, doing a job, which you can do very quickly then come home.
How old will you be? How old will your wife be? How old will your children be?

The answers are 32, 72, 48, 46 and 44!

What research is being done to create a craft that can go at speeds near that of light? In spite of how well humanity has done in space, the thing that keeps us from travel to the stars is the speed of our rockets. They simply aren't fast enough.

Scientists can accelerate particles to speeds approaching that of light in the laboratory but are struggling for spacecraft to even accelerate to more than a tiny fraction of that. Using our current technology, it will take humans about one hundred and fifty days to get to Mars.

All is not lost, though. Among ideas is a system where lasers drive spaceships to Mars in as short a time as three days. This idea depends on the momentum of particles of light (photons) for propulsion. These would come from huge lasers based on Earth and be caught on huge *sails* attached to the craft. Apparently, the technology for this exists and just needs to be greatly enlarged.
At the moment when a spacecraft is launched, the thrust comes from burning rocket fuel. This fuel source weighs down spacecraft and is a very inefficient system in comparison to the

electromagnetic acceleration proposed above. The only limit on electromagnetic acceleration is the speed of light.

Another possible propulsion system is called the *impossible EM Drive*. It has received considerable attention for supposedly providing electromagnetic acceleration, however, scientists are trying to understand how it works, or proves that it can be replicated when required.

The EM Drive does not need any fuel or propellants. It bounces microwave photons back and forth inside a cone-shaped closed metal cavity. Amazingly this causes the pointed end of the EM Drive to push the drive in the opposite direction.

The EM Drive seems to contradict Newton's Third Law but it really does seem to work. Tests are still being done but it does seem promising.

Black Holes, Dark Matter, and Other Phenomena

"There are more things in heaven and earth, Horatio, than is dreamt of in your philosophy."
— Hamlet (1.5.167-8), Hamlet to Horatio

Most people have heard of *black holes* but what are they? Before we can discuss black holes we need to look at what we believe to be the life cycle of a star.

Outlined below are the many steps involved in a star's evolution, from its formation in a nebula, to its death as a white dwarf or neutron star.

Nebula: The womb of a star is a *nebula*. There are different types of nebula, however, we won't dwell on that. At this point, we recall what a star is.

Star: Stars are incandescent balls of gas emitting heat and light as a result of nuclear fusion in their cores. They come from nebulae and have hydrogen and helium gasses as their main constituents.

Temperatures on the surface range from 2000°C to greater than 30,000°C, and their colors vary from red to blue-white. Some stars are very bright and have masses that are 100 times that of the Sun and emit as much light as millions of Suns would. Stars of this sort have lives of less than a million years before they explode as supernovae.

Some stars are very faint; they are called the red dwarfs. They have less than one tenth of one percent of the Sun's brightness.

A star must have a mass at least 8% that of the Sun (that is about 80 times the mass of the planet Jupiter) if it didn't then they could not have a nuclear reaction. If they have less than this critical mass they will only shine dimly. Such stars are called brown dwarves or large planets.

Near the end of its life, a star like the Sun will swell up into a red giant, before it loses its outer layers as a *Planetary Nebula*. It finally shrinks to a white dwarf star.

Red Giant: A red giant is a large bright star with a cool surface. As indicated previously it is created during the latter part of the life of a star like the Sun, as the hydrogen fuel at its center is used up.

The diameters of Red Giants are from 10 to 100 times that of the Sun. Their brightness is caused by their size, even though their surface temperature is less than the surface temperature of the Sun.

Super Giants are very large red giants. Their diameters can be up to 1000 times the diameter of the Sun. Monsters like this can have luminosities 1,000,000 times bigger than the Sun.

Red Dwarf: Red dwarves are small stars that are very cool, faint and, about one-tenth the mass and diameter of the Sun. They are estimated to have lifetimes of 100 billion years as they burn very slowly. Red dwarves include the stars Proxima Centauri and Barnard's Star.

White Dwarf: White Dwarves are very small, hot stars. They are the final stage in the life of a star like the Sun. They have a mass similar to that of the Sun, but they only have 1% of the Sun's diameter. This is about the diameter of the Earth.

White dwarves have a surface temperature of 8000°C or more. However, since they are smaller than the Sun their overall luminosity is no more than 1% of the Sun.

They are the shrunken remains of normal stars, which have used up their nuclear energy supply. They consist of degenerate matter having an extremely high density as a result of gravitational effects. A thumbnail size quantity could have a mass of several tons. It takes over several billion years for a white dwarf to fade away.

Supernova: This is the death throe of a star by an explosion, and frequently results in the star having the brightness of 100 million suns for a short period. There are two types of Supernova but we won't go into this now.

Neutron Stars: These are stars whose main composition are neutrons. They are produced as a result of the explosion of a supernova, which forces the protons and electrons to combine and produce a neutron star. They are very dense.

Most neutron stars have mass three times that of the Sun but a diameter of 20 km or 12 miles. If the mass is greater, the gravity within it is so strong that shrinks further and becomes a black hole.

Pulsars: Pulsars are neutron stars that are spinning very rapidly.

Black Holes: Black holes arise from massive stars at the end of their life cycle. The pull of gravity in a black hole is so vast that nothing, not even light, can escape from its clutches. It is impossible to measure the density of matter in a black hole.

Space around black holes is distorted. They suck any neighboring matter into them, including stars.

Before proceeding, we need to draw attention to a phenomenon called the *Doppler Effect*.

What is the Doppler effect? The Doppler effect is noticed when a wave source has motion with respect to an observer.

In examples of this, there seems to be an apparent shift up in frequency for observers as the source is coming towards the observer and a shift down in frequency as the source is going away from the observer. This is NOT the result of a frequency change, only a wavelength change caused by the movement of the source.

If you have stood on the side of a freeway when there is not much traffic around then you will have probably noticed a good example of the Doppler Effect. A fast moving car approaching you seems to make a different sound than when it passes you.

People who have stood at rail crossings and heard oncoming trains whistle will notice the difference in the sound of the whistle coming from the train as it approaches to when it passes.

Stars emitting light are a wave source and the consequence of the Doppler effect is a difference in the colors of light observed. Some very elegant calculations using this enabled Edwin Hubble to determine that the Universe is expanding.

He also showed there are clusters of galaxies. Some of the clusters have hundreds of thousands of galaxies within them.

This has some extraordinary consequences.

The Big Bang: The Big Bang theory attempts to explain what occurred at the beginning of the universe. Astronomy and physics discoveries have demonstrated that the universe had a beginning. Before then there was nothing and after that our universe. The Big

Bang theory endeavors to explain what transpired during and after that moment.

The Big Bang theory states that the universe came into existence as a *singularity* about 14 billion years ago. You may well ask, "What on Earth is a singularity and where did it come from?" No one knows for certain. Singularities defy our current comprehension of physics. It is believed they exist inside black holes.

After its first appearance, the Big Bang inflated, expanded then cooled. It went from extremely small and extraordinarily hot, to our current universe. It is continuing to expand and cool. This is the Big Bang theory.

The Big Bang was NOT a giant explosion. It was an expansion. It was like a balloon expanding. It was a tiny balloon expanding to be our current universe.

Another misunderstanding is that the singularity was a small fireball appearing somewhere in space. Experts believe space did not exist before the Big Bang. Their calculations show time and space had a finite beginning at the same moment as the origin of matter and energy.

The singularity did not appear in space; space began within the singularity. Before the singularity there was nothing. No space, no time, no matter, no energy; absolutely nothing.

Evidence to support the Big Bang theory?
- the universe had a beginning.

- galaxies are traveling away from us at velocities proportional to the distance from the Earth.

- if the universe at the beginning was very, very hot as the Big Bang suggests, then we should be able to find some remnant of this heat. This has been done
- the large quantities of hydrogen and helium found in the observable universe support the Big Bang model.

Dark Matter: Galaxies perform what looks like an impossible feat. They are spinning at such speed that the gravity created by their observable matter is not capable of holding them together; they should have been flung apart a long time ago. This is also true of galaxies in clusters.

This led scientists to the conclusion that something that cannot be seen is present holding it all together. It is thought there is something yet to be detected which is giving galaxies extra mass to generate the extra gravity required to stay intact. This mysterious substance is called *dark matter* since it is invisible.

In contrast to normal matter, there is no interaction between dark matter and electromagnetic forces. As a consequence it does not reflect, absorb, or emit light. This makes it very hard to observe. The existence of dark matter has been deducted from the gravitational effect it has on visible matter. The mass of dark matter exceeds that of the visible matter by about six to one, making up around 27% of the universe.

There are many theories about dark matter. If one of these theories is true, it will help scientists comprehend the composition of our universe, particularly how galaxies hold together.

There is another substance called *dark energy*. Dark energy is about 68% of the universe and is associated with the vacuum in space. It is distributed evenly in the universe, not only in space but also in time. This means that its effect is not watered down by the expansion of the universe.

The matter that we see and is measurable is only 5% of the universe!

There are other things in outer space that must be mentioned.

Quasars: Quasars are far away objects whose power comes from black holes that are a billion times as massive as the Sun. They shine brighter than galaxies containing them. Quasars have intrigued astronomers since they were discovered 50 years ago.

Cosmic Strings: Cosmic strings are fault lines in the universe, created soon after the Big Bang. There is evidence for their physical existence as quasars show signs of these strings. The mathematics of cosmic string theory is very complicated.

These strings are one-dimensional objects, they have length, but they have no height and they have no width. They are faults in the fabric of the universe, caused by cooling of the universe just after the Big Bang. They can be compared to cracks that form on a frozen lake.

Antimatter: *Antimatter* was one of the most amazing physics discoveries of the 20th century. Science fiction is full of tales of it but people do not realize that it actually exists.

Antimatter is composed of *antiparticles*. Every particle has an antimatter companion that is virtually identical to itself, but with the opposite electrical charge. For an example, an electron has a negative charge. Its antiparticle, named a *positron*, has the same mass but a positive charge. If a particle and its antiparticle meet, there is an annihilation of both resulting in a burst of light.

The British physicist Paul Dirac predicted them as he was attempting to join the two great ideas of early modern physics, which are relativity and quantum mechanics. At the time this seemed an impossibility, as it allowed for negative energies.

Initial skepticism disappeared as examples of these particle-antiparticle pairs were soon discovered. They are produced when cosmic rays enter the atmosphere of the Earth. The energy in thunderstorms seems to produce positrons. Some radioactive decays create them. Work at the Large Hadron Collider (LHC) in Switzerland also produces matter and antimatter, too.

Wormholes: A wormhole is a way of moving through space and time. In theory, it could create a way of greatly shortening long journeys across the universe. The theory of general relativity predicts wormholes. If they actually exist there are dangers of sudden collapse, high radiation, and contact with an exotic matter that could be dangerous.

Albert Einstein with a colleague Nat Rosen, in 1935, proposed the existence of routes through space-time by the use of the theory of relativity. These were called wormholes and would join two different points in space and time, theoretically reducing travel distance and time.

A wormhole has two openings, called *mouths*, with a channel, called *a throat*, joining the two. It is most likely the mouths are spheroidal in shape. The throat could be straight or winding around.

Although the theory of general relativity predicts their existence mathematically none have yet been discovered. A wormhole might be discerned by the way light is affected by gravity in its vicinity.

There are solutions of general relativity allowing for the existence of wormholes with mouths, which are black holes. A black hole occurring naturally, formed by a dying star's collapse, does not create a wormhole by itself.

Wormholes are a favorite of science fiction, which is filled with tales of traveling through them. The reality of such travel would be very complicated, and not just because we haven't found one yet.

Size is the first problem. Naturally occurring wormholes are predicted to be very small, around 10^{-33} centimeters. It is possible, with an expanding universe, that there are wormholes if they exist, which have been stretched.

Stability is another possible problem. The wormholes, which have been predicted, would be useless for traveling in because they collapse quickly. However, there is more recent research showing that a wormhole that contains exotic matter could stay open and constant for longer periods.

Exotic matter is different to dark matter or antimatter. It contains a large negative pressure and negative energy density. Such matter is a part of quantum field theory, the means by which atomic particles are examined.

Theoretically, a wormhole containing enough exotic matter could be used as a means of information being sent. The next step would be sending travelers through space.

It gets even stranger, as wormholes could not only connect two distinct regions within the universe, wormholes could connect two different universes, assuming such exist. There has been conjecture that wormholes could provide a means of time travel. Sadly, the brilliant British scientist Stephen Hawking does not believe this is possible.

Adding exotic matter to a wormhole might make it stable enough for safe human travel along it. The addition of exotic matter might destabilize the portal provided by the wormhole. The technology of today does know how to enlarge or stabilize wormholes; it has not even found one. We are into exciting, unknown territory.

UFOs and Aliens

"For behold, the LORD will come in fire, and his chariots like the whirlwind, to render his anger with fury, and his rebuke with flames of fire"
— Isaiah 6615

This chapter looks at the possibility that as we grope with the ideas and mechanics of travel in space and time we may have been, and are being visited, by the inhabitants of other worlds. This is an idea, which is as old as humanity. The evidence is almost overwhelming that something strange is, and has been occurring.

Given the size of our universe, it seems impossible that we're the only intelligent life. If that's true, then we should see numerous aliens. This is particularly true when the number of stars with planets is considered.

If you have an almost infinite number of stars, then, in turn, that means a vast number of solar systems and planets, and hence an almost certain chance for life, including intelligent life, to evolve.

Calculations give the total number of civilizations at about 1,000; others run to 100,000,000 in the Milky Way alone. There could be 100 million civilizations in the Milky Way galaxy.

Because of the great distances in space and the degradation and distortion of radio signals as they travel long distances we would probably never hear another civilization that used the radio. A good comparison is to ask whether it would be possible for someone to detect the tiny ripple of a stone dropped into the Atlantic Ocean off the coast of New York while standing on a wharf in Europe?

Anyway highly developed civilizations, well beyond ours, would be very unlikely to be using radio waves to communicate. If aliens do not use radio signals to communicate, then the best efforts by organizations like SETI (Search for Extraterrestrial Intelligence) will be futile.

If advanced civilizations are sending messages then they will probably be using technology so advanced beyond ours that we could not possibly pick it up. A similar comparison would be our sending a radio message to a group of monkeys in a jungle tree. The monkeys wouldn't even know we were trying to communicate with them, as they haven't got radios!

As you will see I have no doubt we are not alone; my belief in extraterrestrials is based on personal experience as well as the sheer volume of reports over centuries of visits from other worlds.

Personal experience: In the 1960s I was a university student of mathematics and science. During the long summer holidays, I was living at my parents home in the country and was making money to support myself, at the university, by working at a fellmonger, which is a factory that processes animal hides. I was working very long hours, from 5 AM until 5 PM every day, six days a week.

My parents' home was about 10 miles from where I worked so I had to get up every morning at about 3:30 AM and bicycle to the factory. At that time there had been a number of UFO sightings in that area, which had been written up in the local newspaper. I had read these with some doubts and derision and put them out of my mind.

One morning I got up and was walking to get my bicycle when I saw a very bright light hovering in the sky. I thought it might be a plane or a helicopter but it was not. It stayed where it was for a few minutes then just vanished. No doubt the skeptics would say it was a weather balloon, ball lightning or methane gas. That is the standard cry they make when anyone else relates their sightings of a UFO.

Certainly, the UFO I saw was not very dramatic. One seen by my first wife was. She saw a bright saucer-shaped vehicle land in a wooded area near her father's home on a farm where she was staying.

Many famous people have seen UFOs.

Victoria Beckham: Ms. Beckham is the wife of famous football star, David Beckham, and was a member of the well-known group of pop singers, the Spice Girls. She tweeted that there was a UFO over her home in October of 2011. She also took a photograph, which appeared in the Mirror newspaper.

Elvis: Elvis Presley was one of the most famous singers of all time. Elvis Presley was nearly abducted very early in his career in the 1950s. Together with his crew, which included his bodyguard, Lamar Fike, he was camped out in the desert.

They all looked up and saw a massive, cigar-shaped object hanging in the sky. Elvis said made it an electrical buzzing noise and they all experienced a prickling sensation as the hair on their neck and arms rose.

The presence of the craft caused the bodyguard, Lamar, to collapse face down in the desert sand. Everyone thought that he was dead, his heart had stopped, but they managed to resuscitate him. By that time the UFO had disappeared.

There were several other occasions when the Presley crew saw UFOs, with one actually hovering over the Bel Air mansion of Elvis in California in 1966.

Jimi Hendrix: Jimi was a talented rock guitarist of the 1960s who died of a drug overdose in 1970. On July 30.1970, not long before his untimely death, Hendrix played a concert on the rim of a live volcano in Maui, Hawaii. During the concert a large number of people, Hendricks included, reported seeing UFOs flying overhead.

David Bowie: Bowie was a brilliant British pop singer who passed away in 2016. In 1976 Bowie was in New Mexico filming,' The Man Who Fell To Earth.' Bowie and the film crew had several sightings of UFOs whilst on location. UFOs fascinated Bowie throughout his life. His mother saw one too.

These are but three of the many famous people who claim to have seen UFOs. Go to Google and conduct a search for 'Famous people who have seen UFOs,' to view an ever increasing list.

There are cases involving UFOs where the military has been involved. Two examples of this follow.

From October 1989 throughout 1990, hundreds of reports of illuminated, triangle-shaped objects were recorded in Belgium. On a number of occasions, jets of the Belgian Air Force chased these objects with no success.

Airborne and ground-based radars often saw the objects simultaneously. These sightings remain unexplained, but the Belgian government has cooperated fully with the press and disclosed all details known about these sightings. This, unfortunately, is often not true of other governments.

The most extraordinary case I am aware of occurred during World War 2. It is an absolutely true story.

On the night of 24-5 February 1942, Sirens sounded throughout Los Angeles County. A complete blackout was ordered. Many searchlight beams pierced the black sky.

During the night many anti-aircraft guns fired into the sky at what they believed were enemy aircraft, probably Japanese. Before the night was out, they had fired over 1,400 shells.

The air force was alerted but no planes ever took off. After an hour the guns stopped firing. The blackout order wasn't lifted until daylight.

What was the result? None of the ammunition had hit anything. What was it they had been firing at? Witnesses around Los Angeles had reported a large, dark hovering object, which had been drifting down the California coast.

The local correspondent for the *Herald Express* reported that although the many shells fired at the object were on target, it remained undamaged and untroubled; it eventually moved off at a stately pace between Santa Monica and Long Beach before performing the usual UFO trick of disappearing.

As the defensive action had got under way, observers in Pasadena saw strange red lights on the horizon, moving in erratic patterns.

The report sent by the US Chief of Staff, General George Marshall, to President Roosevelt concluded that no bombs were dropped, no planes were s shot down and no American planes were in action. It stated that "as many as fifteen planes could have been involved, flying at various speeds...from "very slow" to as much as 200 mph and at elevations from 9000 to 18000 feet."

There were no casualties in the armed forces, however, a number of buildings were hit with shells, six civilians were killed, either by the anti-aircraft fire or as a result of the effects of stress arising from the hour-long barrage of artillery fire.

I doubt whether officialdom would try to sweep this under the carpet as weather balloons, methane or ball lightning!

UFOS have been with us for a long time. Prior to the invention of photography, we only had the written testimonies of peculiar events buried within much larger texts and appearing as strange images in medieval woodcuts.

There was one featuring no fewer than 40 UFOs that appeared over Basle in Switzerland on August 7 in 1566. It was found in the Zurich Central library by the great psychologist Carl Jung who devoted much of his later studies to the phenomenon of UFOs.

We could go back half a millennium before the birth of Christ and if we do so we read in the Old Testament of the Bible the peculiar book of Ezekiel. Ezekiel was a Hebrew prophet. He saw something that was just like a UFO and is recorded in the Bible in the book of Ezekiel.

The Egyptians in a scroll dated to around 1500 BC describe a circle of fire in the sky followed by the arrival of others, which landed then took off again.

This is the year 2017. If you Google 'UFO sightings in 2017' you get 675000 hits. Here one of them at http://www.inquisitr.com/3851215/ufo-news-first-2017-sighting-reported/ describes how a British photographer, out for a walk and carrying a high-powered camera, in Devon spotted a mysterious craft flying above. He took a number of photographs of it.

In the same piece, there was a report from Arizona of UFO sightings. The site stated that other reports had come in from the period near the year's end.

There are a huge number of UFO sightings that can easily be explained and the one from Devon probably can, however, there are many which suggest that humanity is not alone in the universe.

Despite what I have just written suppose UFOs exist but are NOT piloted by aliens.

Maybe humans of the future who are traveling in time, possibly through wormholes, or something else that awaits our discovery in the future pilot the UFOs.

The reason why they are having a look at us is the same reason that we would go back and have a look at the dinosaurs and other prehistoric creatures if we found that we could travel in time. These time travelers would not interfere in what is going on now because if they did then their ancestors might not even be born!

The other reason is associated with the lost civilizations of Atlantis, Mu, and Lemuria. A discussion about these will occur in subsequent chapters.

Ancient Civilizations

"When I pronounce the word Future, the first syllable already belongs to the past. "
— Wisława Szymborska

This chapter looks at the possibility that ancient civilizations had contact with extraterrestrials. It looks at some civilizations and events, which lend strength to the suggestion of extraterrestrial contact.

The idea that this happened has been around for a long time but got a huge boost in the 1970s with the publication of the book *Chariots of the Gods* by Erich von Daniken. A feature film was made, based on this book. Many people who had not thought about these ideas before did so. Following the success of von Danniken's book many other books appeared of the same type, some just rehashes of von Danniken's material, but many continuing and expanding on his ideas in a most useful and interesting way.

The Internet became the property of the whole of humanity in the early 1990s, instead of just a medium for academics and scientists. For the last two decades, more and more material of this type has been made available on websites. Some of it is absurd, but other websites offer some absolutely intriguing material.

This chapter will consider some things that point very strongly to extraterrestrial contact.

Piri Reis Map: In 1929 a discovery was made of an amazing map. A Turkish admiral in the sixteenth century called Piri Reis, had it drawn in 1513, copied it from some much older sources.

The map shows the east of South America, the west of Africa, and, most importantly the north of Antarctica. In the map, the coastline of Antarctica is absolutely correct. How was Piri Reis able to draw such a map long before Antarctica region was discovered? To make things even more mysterious the map reveals a coastline, which today, is beneath the ice.

The last period the Antarctic was ice-free ended about 6000 years ago? The ice-free period began sometime between 9000 and 13000 BC!

The Sumerian, Egyptian, Indian and Chinese civilizations appeared after 3500 BC. None of these civilizations could have made that map.

Who or what was here in 4000 years BC or before, able to do things we struggle to do today do with the most powerful technologies?

Deities: The ideas of this chapter suggest that the human population was greatly affected by extraterrestrials visiting Earth in the past. One suggestion is that these aliens involved themselves in the creating of primates, including humans.

Crossbreeding, genetic engineering, or a combination of both did this. This ultimately led to human technologies, cultures, and religion. Many of these ideas include suggestions that deities, including demons and angels, are extraterrestrials whose advanced technology was seized upon by humans as proof of that the aliens were gods.

Some amazing examples of this exist. Here are two.

Bible and sons of God: Genesis 6:1- 22 is an extraordinary passage from the Bible. It says that the sons of God (aliens?) found human women attractive, took them as wives and had children who were mighty men!

A similar piece comes from ancient Sumeria.

Gilgamesh: Gilgamesh was a mythical king known from *The Epic of Gilgamesh* (written in the period 2150-1400 BC), which was the great Sumerian/Babylonian work that predates the *Iliad* of Homer by 1500 years and is the oldest piece of Western literature. In the epic what is described is the interaction between humans and the gods.

This one is a beauty.

Ark of the Covenant: This was the special case that built some 3,000 years ago by the Israelites to store the stone tablets on which were written the Ten Commandments. It was the Throne of God. Biblical accounts describe the building and composition of the Ark in great detail.

It was involved in a number of miracles of the Old Testament. It was carried in front of the Israelites during the Exodus and cleared obstacles from their passage. As soon as the Israelites crossed into the Promised Land across the Jordan River, the Bible says that the river stopped flowing when the Ark-bearers crossed into it.

Later when the Israelites were besieging Jericho, the Ark was carried around the city for a week, with the Israelite soldiers blowing trumpets. On the seventh day, the walls collapsed, allowing them an easy conquest of that city.

Many have asked why the Ark of the Covenant had such powers? Both to help and destroy, and why was the Ark of the Covenant so sacred, was it just because it represented God's throne? Or is there something more to it?

Some people think that the Ark of the Covenant was a very sophisticated device that was capable of many things but dangerous for humans. It has been suggested that the Ark acted as a very large electrical capacitor.

The priests who bore it had to wear special clothing when approaching the Ark; they had to be attired in protective clothing for all important areas of the body.

For many centuries, there has been a vain effort to find and recover the most sacred objects of the Bible. One of the most sought-after is the Ark of the Covenant.

In 597 and 586 B.C., the Babylonians conquered the Israelites, and the Ark disappeared. Was it destroyed? Was it captured? Did someone manage to hide it? These are all unknown.

Let's go to ancient South America.

Tiahuanaco: Tiahuanaco is an abandoned city high in the Andes, two miles above sea level on the shores of Lake Titicaca. It is likely that Tiahuanaco was near sea level once. There are signs of man-made structures below Lake Titicaca. Lake Titicaca is slowly drying up.

It is amazing what the builders of Tiahuanaco did. There are a vast number of precision-cut huge stones at Tiahuanaco (and nearby Puma-punka) moved far from quarries a long way from Tiahuanaco. At such a high altitude or any, this is a remarkable feat. What human did it?

When the Spanish first became aware of Tiahuanaco in the sixteenth century, they tried to destroy it, possibly for religious reasons. Sadly many of the blocks were smashed into material for railway beds early in the last century. The remains of the site are now under protection and are being restored. The ruins of Tiahuanaco could be 14,000 years old!

The most famous archaeological site at Tiahuanaco is the *Sun Gate*. This structure is a calendar. This calendar, though it shows a solar year, does agree with the solar year, as we know it.

The calendar has only 290 days. Some have tried to explain this as a ritualistic calendar. Others have claimed that, 14,000 years ago, that was the length of a year. Still, others think that it may be a year on some other planet. If so then who was the alien who informed the inhabitants?

What discussion of amazing buildings is complete without mentioning pyramids?

Pyramids of Egypt: The idea that aliens from another world may have helped the building of the Great Pyramid of Giza in Egypt could explain some of the most incredible construction in history. The movement of huge stone blocks, quarried at a distance of many miles, then building them into something so enormous, yet exact, seems an impossible feat even today.

Could anyone do it thousands of years ago without modern techniques? The ancient Egyptians must have been onto something or knew someone that we don't.

Mayan pyramids: The Mayan people who lived in southern Mexico built pyramids, starting nearly 3,000 years ago. These are very important tourist attractions for Mexico and Central America.

There were pyramids for sacrificial rituals and pyramids for other sacred ceremonies. The most splendid pyramids of Mesoamerica, as that area is called, were created by the Teotihuacan civilization of the Maya from about 300 B.C. to 500 A.D. There are many sites of pyramids throughout the region.

The pyramids of Teotihuacan have complicated designs and features. The greatest pyramid is the Pyramid of the Sun, which was built on top of a natural cave containing four chambers.

Some pyramids have inner layers of mica brought from Brazil, which is 2,000 miles away. It is believed the mica was some form of insulation. The mica was transported without wheeled transportation.

Ball courts were built at the pyramid of Kukulkan at Chichen Itza. The largest measured 545 feet long by 225 feet wide. There were raised temple areas that improved the acoustics. This allowed voices to be heard at either end of the court. In order to do this, an advanced knowledge of mathematics is necessary.

Around 900 AD, the Maya largely deserted their cities. Some people stayed in the area with their descendants living in Central America to this day. The great Mayan cities fell into ruin, reclaimed by the jungle, without enough people to look after them.

The Maya are called the mysterious Maya, as no one knows where they originated from, or what befell them. It is a great mystery.

Even if aliens helped them build the cities why did the Mayans suddenly leave?

Let's now go to a Pacific Island for more amazing structures which are almost inexplicable without alien input.

Easter Island statues: The mysterious Easter Island statues are on tiny Easter Island in the Pacific. For a long time, it was assumed they were only heads. The statues actually have bodies. Archaeologists have unearthed the bodies of the statues, gradually buried during the 500+ years, which have passed since they were made.

The statues, called *moai* by the Polynesians islanders, were carved out of volcanic rock in a period between A.D. 1100 and 1500. They vary in size, the tallest reaching 10 meters (33 feet). Their purpose is unknown, the moai are believed to be representations of the ancestors of the indigenous peoples. Anthropologists believe they would probably have carved a new statue each time a tribal figure of importance died.

For centuries, anthropologists and other learned folk have wondered how the colossal stone statues of Easter Island moved. It has always been a source of mystery as to how huge statues weighing many tons were transported up to 18 kilometers (11 miles) from the quarry where they were carved, without the help of tractors, cranes, or even large animals.

It was thought that the islanders must have used a combination of ropes, log rollers, and wooden sleds. Now there is a new theory where the *moai,* were moved upright with a rocking motion, using only rope and manpower. If any alien helped move these statues in the past they will be quite amused by this.
In the 1700s the islanders began to write!

Easter Island hieroglyphics: The statues (called moai) are intriguing for their unknown purpose and extraordinary construction, the lost language of *Rongorongo* is equally puzzling. This unique writing seems to have suddenly appeared in the 1700s, but this talent disappeared within two centuries.

Rongorongo was a system of pictographs that was used by the *Rapa Nui,* the inhabitants of the island. It was carved on oblong wooden tablets and on other artifacts. The writing was unknown in any islands nearby and the existence of the script is a great mystery to anthropologists and linguists.

The most logical explanation thus far is that the Easter Islanders were given the idea by the writing they saw in 1770 when the Spanish laid claim to the island. Despite its recent origin, no one has been able to decipher the Rongorongo language successfully.

One who tried has suggested that a sentence with the symbols of a bird, a fish, and the sun says, "All the birds mated with fish: there came forth the Sun." This could be the translation, however, it is totally unlike the chant of an islander who knew its meaning as being about the matings of gods and goddesses. Aliens?

Rongorongo is very interesting to anthropologists, linguists, and archaeologists. Missionaries destroyed as many examples of it as they could find and only twenty-five texts have survived. If anyone finds a meaningful translation of Rongorongo, the surviving tablets could help explain the statues of Easter Island, the sudden appearance of the script, and the history and customs of the island. So far Rongorongo has successfully resisted all attempts at translation.

Atlantis, Mu, and Lemuria: A feature of the legends of all people is the story of a race that lived on this earth, had an extremely advanced technology and had a wonderful civilization. Unfortunately for them, their civilization was filled with violence and wickedness and as a consequence, God, as punishment, destroyed them in a huge flood. Among the cultures and countries that have a flood legend are the Sumerian, the Babylonian, the Irish, the Norse, China, India, the Philippines, and many North and South American native peoples.

In the Bible, this appears as the story of Noah who survives the Flood with his sons Japheth, Shem and Ham and their wives and pairs of every animal in the world. These three sons restarted humanity. In Sumeria, the person who survived the floods and revived life was Atrahasis; in India the person who does this is called Manu.

In South and Central America the people who survived the flood came to be with the people who already lived there. In Mexico it was Quetzalcoatl, in Peru it was Viracocha. These 'gods' brought the fruits of civilization.

It is easy to see that if these people had advanced technology there would be an explanation for the amazing structures throughout the ancient world. If in addition, there were aliens or alien technology among them then there is a ready explanation for the map of Piri Reis.

Conspiracy Theories?

"There's one born every minute."
— PT Barnum

This chapter continues the investigations of the previous chapter and extends them into the possibility that the aliens are already here influencing world events for what are invariably sinister purposes. There are many conspiracy theories and most do not involve extraterrestrials but before we continue it is necessary to say what a conspiracy theory is.

What is a conspiracy theory? The online encyclopedia Wikipedia defines it brilliantly. I recommend you have a look at it for a full definition. Basically, a conspiracy theory is the belief that things are not what they seem, as there are powerful people or forces who are influencing events for their own benefit, gain or power.

The powerful forces, in this case, are aliens and their human collaborators. There are a vast number of websites dealing with a plethora of such theories the first we shall look at is the idea that the Nazis devised flying saucers that may still exist.

Nazis and Flying Saucers: In 1938, or so the legend goes, the Nazis sent a team of scientists to the Queen Maud Land region of Antarctica. During the mapping of this region, they found an extensive system of subterranean warm-water caves and rivers. One of these caves went down 20-30 miles and had a large geothermal lake. This was explored with the result construction teams built a huge base.

The Germans either found alien technology that had been abandoned and made contact with extraterrestrials. They learned how to duplicate alien technology, and used it for a number of advanced weapons including a flying saucer.

Unfortunately for them, and fortunately for us, most of these weapons came too late for use in World War II. There are claims, with some photos to allegedly corroborate this, that they were planning nuclear attacks on both London and New York. There is even a claim they tested a nuclear weapon in the Baltic in late 1944.

If that is the case the question has to be asked why they did not use it on the Soviet troops massing for the final attack on the Fatherland, rather than launch the abortive *Battle of the Bulge*. Those who adhere to this theory worry the base and the ability to make and use these weapons could still exist and some evil group will use it to create a New World Order.

The next conspiracy is that of a small clique of powerful people under such names as *the Bilderberger Group, the Illuminati etc* using or being used by Aliens to bring about World Government: An evil group of people has allegedly always desired to rule the entire world. Military conquest has failed however in our day this conspiracy is employing a discreet but powerful way to achieve this rule. The achievement of this is by dumbing down the education system then gradually using the mass media to get the masses into accepting their ideas. Subtle manipulation, along with distractions (such as the study of useless subjects, unnecessary work, the creation of all sorts of controls, mindless entertainment and obsession with sport) is being deployed. Few are aware of what is going on. These people doing this are not human, they are aliens.

The next conspiracy seems to be the most bizarre yet countless millions are convinced of it.

The Reptilians: the term a former BBC sportswriter, David Icke in his 1999 book The Biggest Secret: The Book That Will Change the World popularized Reptilian. In this book, Icke says that society is under the control of humanoid reptiles. They are called the Reptilians. Although this seems an absurd conspiracy theory, there are about 12 million Americans, and countless others throughout the world, who believe in it.

Members of all Royal Families are supposed to be part of the Reptilian Conspiracy, as are most world leaders. Recently at the beginning of 2017, the Queen did not attend some church services as she normally would. It was claimed that she had demurred about Reptilian plans so she was being kept under house arrest until she ceased her petty rebellion!

Evidence for it is at best anecdotal. Although it may be rubbish, it's a disturbing commentary about how many see how the world works.

Finally, we have the ultimate conspiracy theory.

Satan: The Devil or the slanderer. This is one of the principal titles of Satan, the archenemy of God and of man. It is not known when he originated but there are some clues in the books of Isaiah and Ezekiel in the Bible.

It is certain that he was not created evil. He rebelled against God when in a state of holiness and brought other angels into the rebellion with him. He was cast down with his cohorts to Earth. He is described as a fallen angel. Most sinisterly he is called the god of this world (2Co 4:4).

He is a being of superhuman power and wisdom but he is not all-powerful nor does he know everything. He tries to frustrate God's plans and purposes for human beings. His principal method of attack is by temptation. His power is limited and he can go only as far as God permits him. On the Day of Judgment, he will be cast into Hell to remain there forever.

Closely associated with Satan is the Antichrist. It would seem that the Antichrist would easily be recognized for nothing comparable to him will have been seen before. Many of his activities will be miraculous. His coming will be mysterious. His accomplishments will be such, as the world has never seen before. When the Antichrist actually arrives it would seem that in the beginning almost nobody would recognize him as such. His coming is after all the working of Satan and the world will be completely taken in. Almost any person today who reaches a position of power or popularity is thought to be the Antichrist by some who are gullible. The Bible has many prophecies in it. Many of them have already been fulfilled; some of them still to be fulfilled. There is a very little prophecy concerning the world, the Jews, or the church that can be fulfilled before the coming of the Antichrist.

This is the original conspiracy theory and it can easily be interpreted in such a way that aliens from other worlds or galaxies are the angels. All religions feature the struggle between good and evil.

The part of the previous chapter that talked about deities is very relevant here.

Final Thoughts

"The future belongs to those who believe in the beauty of their dreams."
— Eleanor Roosevelt

We are fortunate to live in a time when it is possible to discuss how to travel in Space and Time could be done.

For all of us, we can realistically look forward to a time when our grandchildren can go to the stars. I believe that if you take the time to grasp what this book introduces then you will help get them there.

Space and Time are vast fascinating subjects.

Knowledge acquired about them has been built up over many centuries

This knowledge has led to technologies that are at the pinnacle of human achievement.

It is possible this knowledge was known in the past, imparted by those who had mastered travel through space and time.

It would be tragic if this knowledge and technology were misused!

Good luck to you in your journey!

www.ingramcontent.com/pod-product-compliance
Lightning Source LLC
Chambersburg PA
CBHW061216180526
45170CB00003B/1017